Physics
Demystified

(100 PHYSICS II)

Simple Problems to Boost Your Skills

Alistair Jonas

Table of Content

Heat Energy

Heat or thermal is a form of energy due to temperature changes in a substance, object or place. Heat energy sources include; the sun, friction, chemical reactions and burning.

Example 1

How much heat is given when a piece of iron of mass 50g and specific heat capacity of 460J/kgk cools from 85°C to 25°C

<u>Solution</u>

Heat $Q = mc\Delta\emptyset$

$c = 460J/kgk$

$m = 50g = 50/1000 = 0.05kg$

$\Delta\emptyset = 85 - 25 = 60°C$

$Q = 0.05 \times 460 \times 60 = 1380J$

$= 1.38 \times 10^3 J$

Example 2

How much heat is emitted wihen a body of mass 200 g cools from 37°C to 31°C? (specific heat capacity of the body 0.4Jg-^{1}k-1).

Solution

Quantity of heat $Q = mc\Delta\emptyset$

$m = 200g$

N.B. we convert m to kg only if c is in Jkg-^{1}k-^{1}

$\Delta\emptyset. = [37 - 31] = °6C$

$Q = 200 \times 0.4 \times 6 = 480J$

Example 3

A piece of substance of specific heat capacity 450Jkg'k falls through a vertical distance of 20m from rest. What will be the rise in temperature of the substance on hitting the ground when all its energies are converted into heat.

Solution

quantity of heat $= mc\Delta\varnothing$

Where m= mass in kg

SHC specific heat capacity=c in $JKg^{-1}k^{-1}$

Temperature change $= \Delta\varnothing$

Recall heat is a form of energy= mgh

where m.= mass g = acceleration due to gravity h= height

$\therefore \quad mgh = mc\Delta\varnothing$

$m \times 10 \times 20 = mc\Delta\varnothing$

divide through by m

$10 \times 20 = 450 \times \Delta\varnothing$

$200 = 450 \times \Delta\varnothing$

$\Delta\varnothing = 200/450 = 0.44\,°C$

Example 4

Heat is supplied uniformly at the rate of 100W to 1.0×10^{-2}Kg of a liquid for 20 seconds. If the temperature of the liquid rises by 5°C then the specific heat capacity of the liquid is?

<u>Solution</u>

Heat supplied = power x time

t = 20s

Power = 100w

Heat supplied = 100 x 20 = 2000J

Heat gained by liquids = mcΔØ

C =? ΔØ = 5°C. m = 1.0×10^{-2}Kg

heat gained by liquid = 1.0×10^{-2} x c x 5

heat gained = heat supplied.

1.0×10^{-2} x c x 5 = 2000

$c = \dfrac{2000}{1.0 \times 10^{-2} \times 5}$

c = 4×10^4 JKg^{-1}k^{-1}

Example 5

A blacksmith cools an iron bolt mass 0.5kg temperature 400°C by putting it into a pail containing 9kg water at 20°c. Find the final temperature of the water and bolt. Ignore the heat gained by the pail itself and any steam which may be emitted. (s.h.c. of water & iron are 4200J/kgk and 450J/kg'krespectively)

Solution

Let mass of iron be m_i = 0.5kg

SHC of iron C_i = 450J/kg k

Temperature of iron $\emptyset_i$ = 400°c

Final temperature $= \emptyset_f$

mass of water m = 9kg

Temp of water = $\emptyset_w$ = 20°c

SHC of water = C_w

Heat gain by water = $mC_w(\emptyset_f - \emptyset_w)$

Heat lost by the iron = $m_iC_i(\emptyset_i - \emptyset_f)$

Heat lost by iron = Heat gained by water

∴ $0.5 \times 450(400 - Ø_f) = 9 \times 4200 \times (Ø_f - 20)$

Divide both side by 9×25

$(400 - Ø_f) = 168(Ø_f - 20)$

Open the bracket and collect like terms

$400 - Ø_f = 168Ø_f - 3360$

$400 + 3360 = 168Ø_f + Ø_f$

$3760 = 169Ø_f$

$Ø_f = \dfrac{3760}{169}$

$= 22.2°C$

final temperature $= 22.2°C$

Example 6

A given quantity of heat increases the temperature of 150 grams of water from 9C to 15°C and increases the temperature of an equal volume of oil weighing 100 grams from

9°C to 25°C. The ratio of the SHC of oil to that of water is?

<u>Solution</u>

Let the specific heat capacity of water be Cw and that of the oil be Ci then

$$mC_w\Delta\phi_w = mC_i\Delta\phi_i$$

mass of water $m_w = 150g$

mass of oil $m_i = 100g$

$\Delta\phi_w = (15 - 9) = 6$

$\Delta\phi_i = 25 - 9 = 17$

$150 \times C \times 6 = 100 \times C_2 \times 16$

divided through by 50

$A\phi = (25 - 9°c$

$3 \times C_w \times 6 = 2 \times C_i \times 16$

$$\frac{C_w}{C_i} = \frac{3 \times 6}{2 \times 16}$$

$$= 0.56$$

Example 7

How long does it take a 750W heater to raise the temperature of lkg of water from 20°C to 50°C?

Solution

Specific heat capacity of water 4200J/KgK

heat supplied = power x time

Power = 750W

t = ?

heat supplied = 750 x t

Heat gained by water = $mC\Delta\emptyset$

m = lkg

$\Delta\emptyset$ = (50- 20)'C

c = 4200Jkg K

heat gained = 1x 4200 x (50 - 20)

= 30 × 4200

heat supplied = Heat gained by water

750 x t= 30 x 4200

t= 30 x 4200

$$\frac{750}{}$$

= 168 second or 2.8min

Example 8

A piece of copper ball of mass 20g at 200C is placed in a copper calorimeter of mass 60g containing 50g of water at 30°C. Ignoring heat loses, Calculate the final steady temperature of the mixture

Specific heat capacity of water =4.2J/g K.

Specific heat capacity of copper =0.4J/gK.

Solution

Heat lost by copper ball = Heat gained by Calorimeter + heat gained by water.

$$mCu(\emptyset^1 - \emptyset) = McCc[\emptyset - \emptyset^{11}] + MwCw [\emptyset - \emptyset^{11}]$$

mass of copper ball m = 20g

Temperature of copper $\emptyset^1$ = 200°C

Mass of copper colorimeter Mc = 60g

Mass of water Mw = 50g

initial temperature of water $\emptyset^{11} = 30°C$

final temperature of mixture $\emptyset =$?

Specific heat capacity of copper $Cu = 0.4 J/gk$

Specific heat capacity of calorimeter $Cc = 0.4 J/gk$

$$20 \times 0.4(200 - \emptyset) = 60 \times 0.4 (\emptyset - 30) + 50 \times 4.2(\emptyset - 30)$$

$$8 (200 - \emptyset) = 24 (\emptyset - 30) + 210 (\emptyset - 30)$$

$$1600 - 8\emptyset = 24\emptyset - 720 + 210\emptyset - 6300$$

$$1600 + 720 + 6300 = 24\emptyset + 8\emptyset + 210\emptyset$$

$$8620 = 242\emptyset$$

$$\emptyset = \frac{8620}{242}$$

$$= 35.6°K$$

Latent Heat

Example 9

How much heat is required to convert 20g of ice at 0'C to water at the same temperature (specific latent heat of ice 336Jg^{-1})

<u>Solution</u>

<u>N.B</u> :

Latent heat is the heat requires to cause a change of state of matter. This change of state could be fusion or vaporization

Latent heat is independent on temperature

Therefore latent heat $Q = ml$

Where m mass = 20g

l = latent heat= 336J/g

$Q = 20 \times 336$

$= 6720J$

Example 10

Calculate the heat required to convert ice at 0°C to water at 16°c (Specific latent heat of fusion ice 336Jg] (Specific heat capacity of water 4.21g^{-1}K^{-1}

<u>Solution</u>

Total heat required = latent heat of fusion + heat gain by ice

$$Q = ml + mC\Delta\emptyset$$

$$m = 20g$$

L=336J/g. SHC of water C = 4.21

Temperature change $\Delta\emptyset = (16 - 0)$

$$Q = 20 \times 336 + 20 \times 4.2 (16 - 0)$$

$$Q = 6720 + 1344$$

$$= 8064J$$

$$= 8.064 \times 10^3 J$$

Example 11

Calculate the mass of ice that would melt when 2Kg of copper is quickly transferred from boiling water to a block of ice without heat loss. [Specific heat capacity of copper $M_c = 400 J/gK$]. [SLH of fusion of ice $L = 3.3 \times 10^5 J/kg$]

Solution

heat lost by copper = heat gained by ice

mass of copper M_c = 2kg

$\Delta\emptyset$ temperature change of ice from 0 to boil=100° c

Mass of ice m

heat lost by copper = heat gained by ice

$\therefore M_c C_c \Delta\emptyset = mL$

$2 \times 400 \times 100 = m \times 3.3 \times 10^5$

$$m = \frac{80000}{33 \times 10^4}$$

$$= 0.24kg$$

Example 12

An electric thermal machine is used to melt 1.5Kg of ice. If the machine is powered by a 12V, 20A battery, calculate the time taken to melt the ice at 0°C [Specific latent heat of fusion of ice 336 x 10³J/kg]

Solution

Let t be the time in second taken to melt the block of ice

heat supplied = Ivt

heat absorbed to 0°C = mL

V = voltage = 12v

t = ?

I = 20A

m = 1.5Kg

L =

heat supplied = heat gained or absorbed

Ivt = ml

20 x 12 xt = 1.5 x 336 x 10³

$$t = \frac{1.5 \times 336 \times 10^3}{20 \times 12}$$

$$t = 2100 \text{ second}$$

Example 13

0.5Kg of water at 10°C is completely converted to ice at 0°C by extracting 188000J of heat from it. If the specific

heat capacity of water 4200J/kg C, calculate the latent heat of fusion of ice.

Solution

$$Q = mc\Delta\emptyset + mL$$

mass (m)= 0.5Kg

L=?

$$\Delta\emptyset = (10-0) = 10$$

188000 = 0.5 x 4200 x 10 +0.5L

188000 = 21000+ 0.5L

188000- 21000= 0.5L

167000= 0.5L

$$L = \frac{167000}{0.5}$$

$$= 334{,}000 \text{J/kg}$$

$$= 3.3 \times 10^5 \text{J/kg}$$

Example 14

A refrigerator converts 500g of water at 20'C into ice at -10C in 2 hours. What is the rate at which heat energy is extracted from the water? [Specific latent heat of fusion of water = 336000JKg, specific

heat capacity of water = 4200JKg K

specific heat capacity of ice 2100JKg 'K')

Solution

Heat energy extracted in cooling the water from 20°C to 0°C.

$$Q^1 = mCw\Delta\varnothing$$

$$m = 0.5\text{Kg} \quad Cw = 4200\text{JKg'K}$$

$$\Delta\varnothing = 20 - 0 = 20$$

18

$Q1 = 0.5 \times 4200 \times 20 = 42000J$

Heat energy extracted when the water changes to ice Q2

$Q2 = ml$

$m = 0.5kg$ $L = 336000jkg$

$Q2 = 0.5 \times 336000 = 168000J$

Heat energy extracted when the ice cools from 0°C to -10°C

$Q3 = mc\Delta\emptyset$

$m = 0.5$ $c = 2100JKg\ K$

$\Delta\emptyset = (0-(-10)$

$= 10°C$

$Q3 = mc\Delta\emptyset$

$= 0.5 \times 2100 \times 10$

$= 10500J$

Total heat energy $= Q1 + Q2 + Q3$

$42000 + 168000 + 10500$

$= 220500 = 220.5KJ$

Example 15

What is the difference in the amount of heat given out by 4Kg of steam and 4Kg of water when both are cooled from 100°C to 80'C? [Specific latent heat of steam = 2,260000 (Specific heat capacity of water 4,200jkg']

<u>Solution</u>

Heat give out by steam Q1 = mL

L = 2,260000J/kg

Q = 4 x 2,260000J

Heat loss by water I00°C to 80°C Q2 = $mc\Delta\emptyset$

m = 4kg

SHC of water C = 4200

Q2 = 4x 4,200 x (100 - 80)

= 4 x 4.200 x 20

difference

Q1 - Q2 = (4 x 2,260000) - (4, 4,200 x 20)

= 9040000 - 336000

$$= 8,704,000J$$

Example 16

An electric kettle rated at 1500W boils away 0.3kg of a liquid at the boiling point of water in 300s. Calculate the specific latent heat of vaporization of the liquid.

Solution

Electrical energy supplied. = power x time

$$pt = mL$$
$$P \ 1500W$$
$$m = 0.3 \ kg$$
$$t = 300s$$
$$1500 \ x \ 300 = 0.3 \times L$$
$$L = \frac{1500 \times 300}{0.3}$$
$$= 1500000J/kg$$
$$= 1.5 \times 10^6 J \ kg^{-1}$$

Example 17

A piece of copper of mass 300g at a temperature of 950°C is quickly transferred to a vessel of negligible thermal capacity containing 250g of water at 25°c. If the final steady temperature of the mixture is 100°C. Calculate the mass of the water that will boil away.

Specific heat capacity of copper = 4.0 x 10°kg'k Specific heat capacity ot water= 4.2 x 10°kg'k!

Specific latent heat of vaporization of steam = 2.26 x 10⁶Jkg

m = mass of steam that boil away

$$Mw = 250g$$

$$\Delta\varnothing w = (100 - 25)c$$

$$0.3 \times 400 \times (950 - 100) = m \times 2.26 \times 10^{\circ} + 0.25$$

$$\times 4200 \times (100 - 25)$$

$$102000 = m \times 2.26 \times 10^{\circ} + 78780$$

$$102000 - 78750 = 2.26 \times 10^{6}m,$$

$$23250 = 2.25 \times 10^{6}m$$

$$m = \frac{23250}{2.25 \times 10^{6}}$$

$$= 0.0103kg$$

$$= 10.3g$$

Example 18

A mass of a liquid at 30°C is mixed with a mass of the same liquid at 70°C and the temperature of the mixture is 45°C. Find the ratio of the mass of the cold liquid to the mass of the other liquid.

Solution

$$M_1 C \Delta\emptyset_1 = M_2 C \Delta\emptyset_2$$

$$\Delta\emptyset_1 = (45 - 30) = 15$$

$$\Delta\emptyset_2 = (70 - 45) = 25$$

$$M_1 \times C \times 15 = M_2 \times C \times 25$$

Divide through by C

$$M_1 \times 15 = M_2 \times 25$$

$$\frac{M_1}{M_2} = \frac{25}{15}$$

$$= 5/3 = 5 : 3$$

Example 19.

How many grammes of water at 17°C must be added to 42g of ice at 0°C to melt the ice completely [Specific latent heat of fusion of ice 3.4×10^5 Jkg, specific heat capacity of water = 4200J/kgk. Ans =200g

Solution

M_w = mass of water = ?

M_i = mass of ice = 42g

$\Delta\varnothing_w$ = 17

C_w = SHC of water = 4200J/kg/K

C_i = Specific latent heat of ice = 3.4×10^5

Heat lost by water = $M_w C_w \Delta\varnothing_w$

$= M_w \times 4200 \times 17$

Heat gain by ice = $M_i C_i$

$= 42 \times 3.4 \times 10^5$

Heat lost = heat gain

$M_w \times 4200 \times 17 = 42 \times 3.4 \times 10^5$

$M_w = \dfrac{42 \times 3.4 \times 10^5}{4200 \times 17}$

$= 200g$

Example 20

What is the energy rate of a heater used to supply energy for 3 minutes. The energy supplied is used to completely turn a liquid of mass 6kg into vapour. Specific latent heat of vaporization of liquid = $2.26 \times 10^3 Jkg^{-1}$

Solution

P = ?

t = 3min = 3 × 60 sec

Heat energy supplied = P × t

= P × 3 × 60 = 180

latent heat of vaporization L = $2.26 \times 10^3 Jkg^{-1}$

Heat gained by liquid M × L

= $6 \times 2.26 \times 10^3 = 13.56 \times 10^3$

Heat supplied = heat gained by liquid

P × 180 = 13.56×10^3

$P = \dfrac{13.56 \times 10^3}{180}$

= 75.33W

Exercises on Heat Energy, SHC and Latent Heat

1. How much heat is given out when a piece of brass of mass 150g and specific heat capacity 560J/kgk cools from 90'c to 35'C. Ans= 4.62KJ

2. A piece of iron of mass 120g and specific heat capacity 350J/kgk cools from 80°C to 25°C. The heat released by the iron is? Ans = 2310J

   A piece of metal of mass 140g and specific heat capacity 400Jkg^{-1}k^{-1} loses 440J of heat energy. The change in temperature of the body is? Ans = 7.9°C

3. A body of substance which has 400Jkg^{-1}k^{-1} as specific heat capacity falls through a vertical distance of 30m from rest. Calculate the rise in temperature of the

substance on hitting the ground when all its energies are converted to heat.

{Hint : mgh = mc$\Delta\emptyset$}   Ans = 0.75°C

4. 500g of water is heated for 7 minutes so that its temperature rises from 30°C o 72°C.   Calculate the heat supplied per minute. [Specific heat capacity of water 4200 J Kg^{-1}k^{-1}] Ans =12600J

5. Calculate the time taken to heat 2kg of water from 50C to 100°C in an electric kettle taking 5A from a 210V supply. [Specific heat capacity of water 4200Jkg] Ans = 800s

6. A Hydroelectric kettle with negligible heat capacity is rated at 2000W. If 2.0kg of water is put in it, how long will it take the temperature of water to rise from 20°C to 100°C [Specific heat capacity of water = 4200Jkg'k] Ans = 336sec

7. A heater coil supplies 1000W which is used to boil off 0.5kg of boiling water. The time taken to boil off the water is Specific latent heat of vaporization is 2.3×10^6 T/kg. Ans = 11.5×10^2 sec

8. 250g of lead at 170C is dropped into 100g of water at 0°c. If the final temperature is 12°C, the specific heat of lead is? [Specific heat capacity of water is 4200Jkg°c]. Ans= 127.6Jkg^{-1}K^{-1}

9. All the heat generated by a curent of 2A passing through a 6 ohms resistor for 25s is used to evaporate 5g of a liquid at its boiling point. What is the specific latent heat of the liquid? Ans = 120Jg^{-1}K^{-1}

10. An immersion machine is used to supply energy of 90 watts for five minutes. The energy supplied is used to completely melt 180g of a solid at its melting point.

Neglect energy loss to the surroundings, calculate the specific latent heat of fusion of the solid. Ans = 1.5J/kg or 150J/g

11. A heater marked 50W will evaporate 0.005kg of boiling water in 50seconds. The specific latent heat of vaporization of water in Jkg^{-1} is? Ans = {5×10^5J/kg

12. A 500W heater is used to heat 0.6kg of water from 25°C to 100°C in t_1 seconds, If another 1000W heater is used to heat 0.2kg of water from 10 to 100°C in t_2 seconds. Find t_1 : t_2 Ans = 5:1

13. A body with a 450Jkg-1k-1 specific heat capacity drops 20 meters to the ground from rest. Calculate the change in ground temperature caused by the body colliding with the ground, assuming energy conservation. Ans = 4/9°C

14. When two objects P and Q are supplied with the same quantity of heat, the temperature change is observed to be twice that in Q. If the masses of P and Q are the same. Calculate the ratio of the specific heat capacities of Q to P. Ans = 1:2

15. A 0.5kg liquid is heated with a 400W immersion heater. if the liquid's temperature rises by 2.5°C. Calculate the liquid's specific heat capacity in one second while ignoring heat losses to the environment. The mixture's temperature is 50° C. A = 320 J/kg K

16. An electric kettle rated at 2500W boils away 0.4kg of a liquid at its boiling point in 250 second. Calculate the specific latent heat of vaporization of the liquid. Ans = 1.56×10^6

17. An electric kettle rated at 1200W boils away 0.5kg of a liquid at its boiling point in 200seconds. Calculate specific latent heat of vaporization of the liquid. Ans = 4.8×10^5J/kg

18. A piece of copper of mass 250g at a temperature of 650°C is quickly transfer to a vessel of negligible thermal capacity containing 150g of water at 20°C. If the final steady temperature of the mixture is 100°C. Calculate the mass of the water that will boil away specific heat capacity of copper = 5.0×10^2Jkg' k specific heat capacity of water 4.2×10^3J/kgk. S.L.H of vaporization of steam =2.26×10^6Jkg^{-1}
Ans= 8.1g

19. A glass containing 500g of squash at 20°C has 100g of ice at 0°C put into it What will be the temperature of the squash

when the ice has all melted? Specific latent heat of fusion of ice $=336000Jkg^{-1}$ specific heat capacity of water $4200Jkg^{-1}k^{-1}$ Assume squash has the same specific heat capacity as water. Ans = 3.33°C

20. A mass of a liquid at 60°C is mixed with a mass of the same liquid at 90°C and the temperature of the mixture is 30°C. Find the ratio of the mass of the cold liquid to the mass of the other liquid. Ans = 2:1

Temperature

Example 21

The lowest fixed point of a temperature scale is 40 mm, while the upper fixed point is 200 mm. What does a thermometer reading of 60 °C mean on this scale?

Solution

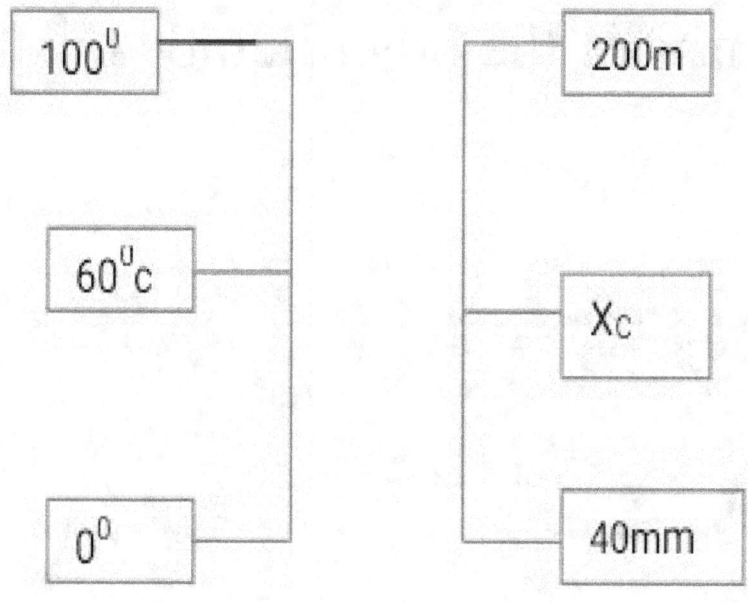

N.B

The idea is to compare the mm scale with the celsius scale.

L100 = upper fixed point of celcius scale or centigrade scale = 100°c

L60 = middle scale 60°c

L0 = lower fixed point = 0

$$\frac{L60 - L0}{L100 - L0} = \frac{Xc - X40}{X200 - X40}$$

$$\frac{60 - 0}{100 - 0} = \frac{Xc - 40}{200 - 40}$$

$$\frac{60}{100} = \frac{Xc - 40}{160}$$

$$\frac{60 \times 160}{100} = Xc - 40$$

$$96 = Xc - 40$$

$$Xc = 96 + 40$$

$$Xc = 136mm$$

Example 22

A thermometer has lower and upper fixed points as 6cm and 48cm what are the temperature when Ø are (i) 1 5cm (ii) 30cm

Solution

N.B

You may sketch the two type of thermometer as above before comparing the upper, lower and unknown point.

i. The upper fixed points L_{100} at 48cm = 100

Lower fixed point L_0 at 6cm = 0

Temperature at 15cm L_{15} = ?

X represents Ø

$$\frac{L_{100} - L_0}{L_{15} - L_0} = \frac{X_{48} - X_6}{X_{15} - X_6}$$

$$\frac{100 - 0}{L_{15} - 0} = \frac{48 - 6}{15 - 6}$$

$$\frac{100}{L_{15}} = \frac{42}{9}$$

Cross multiply

$$42L_{15} = 900$$

$$L_{15} = 900/42$$

$$= 21.4^{0}C$$

ii. The upper fixed points L_{100} at 48cm $= 100$

Lower fixed point L_0 at 6cm $= 0$

Temperature at 30cm $L_{15} = ?$

$$\frac{L_{100} - L_0}{L_{30} - L_0.} = \frac{X_{48} - X_6}{X_{30} - X_6}$$

$$\frac{100 - 0}{L_{30} - 0} = \frac{48 - 6}{30 - 6}$$

$$\frac{100}{L_{30}} = \frac{42}{24}$$

Cross multiply

$$42L_{15} = 2400$$

$$L_{15} = 2400/42$$

$$= 57.14^{0}C$$

Example 23

A thermometer has its stem marked in millimeter instead of degree centigrade. The lower and stream point are 15mm and 120mm respectively. calculate temperature in çentigrade degree when the thermometer read 35mm.

Solution

L represent centigrade scale while Ø represent mm scale

$$\frac{L_{100} - L_0}{L_{35} - L_0} = \frac{Ø_{120} - Ø_{15}}{Ø_{35} - X_{15}}$$

$$\frac{100 - 0}{L_{35} - 0} = \frac{120 - 15}{35 - 15}$$

$$\frac{100}{L_{35}} = \frac{105}{20}$$

Cross multiply

$$105L_{15} = 2000$$
$$L_{15} = 2000/105$$
$$= 19.05^0C$$

Example 24

The lowest fixed point on a temperature scale is 20 mm, while the upper fixed point is 180 mm. When the scale reads 70 °C, determine the temperature.

Solution

Let the Celsius scale be L and the mm scale be X

Upper fixed points $L_{100} = 100$ when $X_{180} = 180mm$

Lower fixed point $L_0 = 0$ when $X_{20} = 20mm$

When celsius scale $L_{70} = 70$, $X_? = X$

$$\frac{L_{100} - L_0}{L_{70} - L_0.} = \frac{X_{180} - X_{20}}{X_? - X_{20}}$$

$$\frac{100 - 0}{70 - 0} = \frac{180 - 20}{X - 20}$$

$$\frac{100}{70} = \frac{160}{X - 20}$$

Cross multiply

$$100(X - 20) = 70 \times 160$$

$$100X - 2000 = 11200$$

$$100X = 11200 + 2000$$

$$100X = 13200$$

$$X = \frac{13200}{100}$$

$$= 132mm$$

Example 25

The lower and upper fixed ed points of a mercury in glass thermometer are marked y and 180mm respectively on a particular day the mercury meniscus in the thermometer rises to 45mm and the corresponding reading on a Celsius scale 10°c, calculate the value of y.

Solution

Let the Celsius scale be L and the mm scale be X

Upper fixed points $L_{100} = 100$ when $X_{180} = 180mm$

Lower fixed point $L_0 = 0$ when $X_y = y$

When celsius scale $L_{10} = 10°c$, $X_{45} = 45$

$$\frac{L_{100} - L_0}{L_{10} - L_0} = \frac{X_{180} - X_y}{X_{45} - X_y}$$

$$\frac{100 - 0}{10 - 0} = \frac{180 - y}{45 - y}$$

$$\frac{100}{10} = \frac{180 - y}{45 - y}$$

Cross multiply

$$100(45 - y) = 10(180 - y)$$

$$4500 - 100y = 1800 - 10y$$

Collect like terms

$$4500 - 1800 = -10y + 100y$$
$$2700 = 90y$$
$$y = \frac{2700}{90}$$
$$= 30mm$$

Example 26

A mercury in glass thermometer reads -25° at ice point and 120° at steam point. Calculate the Celsius tempeature corresponding to 60 on the thermometer.

Solution

Upper fixed points $L_{100} = 100$ when $X_{120} = 120$

Lower fixed point $L_0 = 0$ when $X = -25$

When celsius scale $L? = L$ $X_{60} = 60$

$$\frac{L_{100} - L_0}{L_? - L_0.} = \frac{X_{120} - X}{X_{60} - X}$$

$$\frac{100-0}{L} = \frac{120-(-25)}{60-(-25)}$$

$$\frac{100}{L} = \frac{145}{85}$$

Cross multiply

$$145L = 8500$$

$$L = \frac{8500}{145}$$

$$= 58.62°C$$

Temperature conversion from Celsius °C to Kelvin°K to Fahrenheit °F

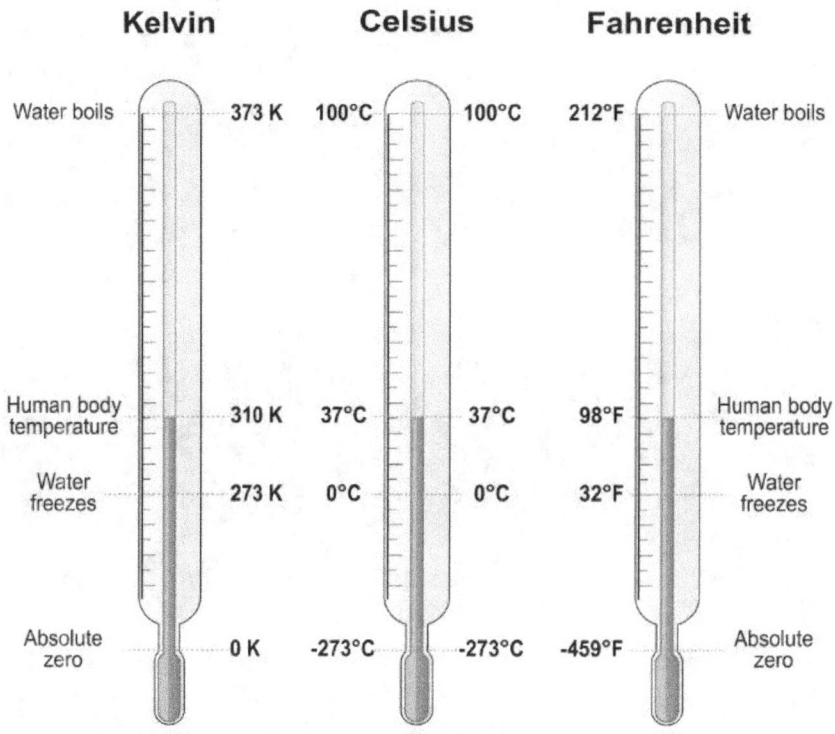

The formulae for temperature conversion is as follows

- From ° C to °K

  °K= ° C + 273

Absolute Kelvin scale has upper fixed points as 373 K and lower fixed point = 273K

Example 27

The melting point of naphthalene is 78°c. What is this temperature in Kelvin?

Solution

$$Kelvin = 0°c + 273$$
$$= 78° + 273$$
$$= 351k$$

• From °K to °C

$$°C = °K - 273$$

Example 28

What is the celsius equivalent of 370°K

Solution

$$°C = K - 273$$
$$= 370 - 273$$

$$= 97\ ^{\circ}C$$

- From $^{\circ}$C to $^{\circ}$F

  $^{\circ}F = 1.8\ ^{\circ}C + 32$

  The upper and lower fixed point of a Fahrenheit scale are 212°F and 32°F respectively

Example 29

A temperature of 30 $^{\circ}$C has an equivalent Fahrenheit scale as?

Solution

$$^{\circ}F = 1.8\ ^{\circ}C + 32$$
$$= 1.8 \times 30 + 32$$
$$= 54 + 32$$
$$= 86\ ^{\circ}F$$

- From °F to °C

  $$^{\circ}C = \frac{5(^{\circ}F - 32)}{9}$$

Example 30
Convert -40 ° F To °C

Solution

$$°C = \frac{5(F - 32)}{9}$$

$$= \frac{5(-40 - 32)}{9}$$

$$= \frac{5(-72)}{9}$$

$$= 5(-8) = -40°C$$

Example 31
Convert 104°F to Kelvin

Solution

First convert 104F to °C

$$°C = \frac{5(104 - 32)}{9}$$

$$= \frac{5(72)}{9}$$

$$= 40°C$$
$$\text{Kelvin } K = °C + 273$$
$$= 40 + 273$$
$$= 313K$$

Example 32
Convert 273 K to Celsius

Solution

First convert 273K to Celsius

$$C = K - 273$$
$$= 273 - 273 = 0 \, °C$$
$$F = 1.8°C + 32$$
$$= 1.8 \, (0) + 32$$
$$= 32$$

Exercises on Temperature and Temperature conversion

1. The ice and steam point on a mercury in glass thermometer are found to be 85.0mm apart. At what temperature in Celsius degree will the length of the mercury thread be 31mm above the ice point mark. Ans = 36.47°C

2. Ice and steam have differences in melting points of 150 mm on a mercury-in-glass thermometer. Calculate the temperature in degree Celsius that is seen when the mercury thread is 35.6mm above the ice point.
   Ans = 23.73°C

3. The ice and stream point of mercury in glass thermometer reads -30° and 100. Calculate the Celsius temperature

corresponding to 85° on the thermometer. Ans = 88.4-°C

4. A faulty mercury in glass thermometer has 1.8 C and 105.0°C as its ice and steam point respectively. When the the temperature is 50°C the reading on the, the thermometer is? Ans = 53.4°C

5. A platinum resistance thermometer read 12.4ohms and 20.4ohms lower and upper fixed points respectively. Calculate' the temperature when the resistance is 18.5 ohms. What will be the resistance of the wire when the temperature read 50°c. Ans = 76.25°c , 16.14ohms

www.ingramcontent.com/pod-product-compliance
Lightning Source LLC
Chambersburg PA
CBHW081546220526
45467CB00010B/3345